Simple Space Rockets Colouring, Drawing & Activity Book

Written by: L k D

Series 2

Simple Space Rockets: Colouring, Drawing & Activity Book by L.K.D

Attribution of Simple Space Rockets: Colouring, Drawing & Activity Book to Vecteezy.com.

Published by L.K.D

Toronto, Ontario, Canada.

Copyright © 2021 L.K.D

All rights reserved. No portion of this book may be reproduced in any form, stored in any retrieval system, or transmitted in any form by any means---electronic, mechanical, photocopy, recording, or otherwise without permission from the publisher, except as permitted by copyright law. For permissions contact:

Email: info@kidspill.com

Cover & Illustration by L.K.D

Website: www.kidspill.com

This book belongs to:

Draw the spaceship

Are all the rockets different?

YES   NO

FIND THE WORDS:
ALIEN
SPACESUIT
HELMET
SATALLITE
STARS
ASTRONAUT
LAUNCH

```
T U A N O R T S A P
E E T I L L A T A S
B S T E M L E H L Z
Q D W H A S X V J Y
F G Y L Z Y R V U F
H G I T E K U A R U
U E L A U N C H T G
N T I U S E C A P S
```

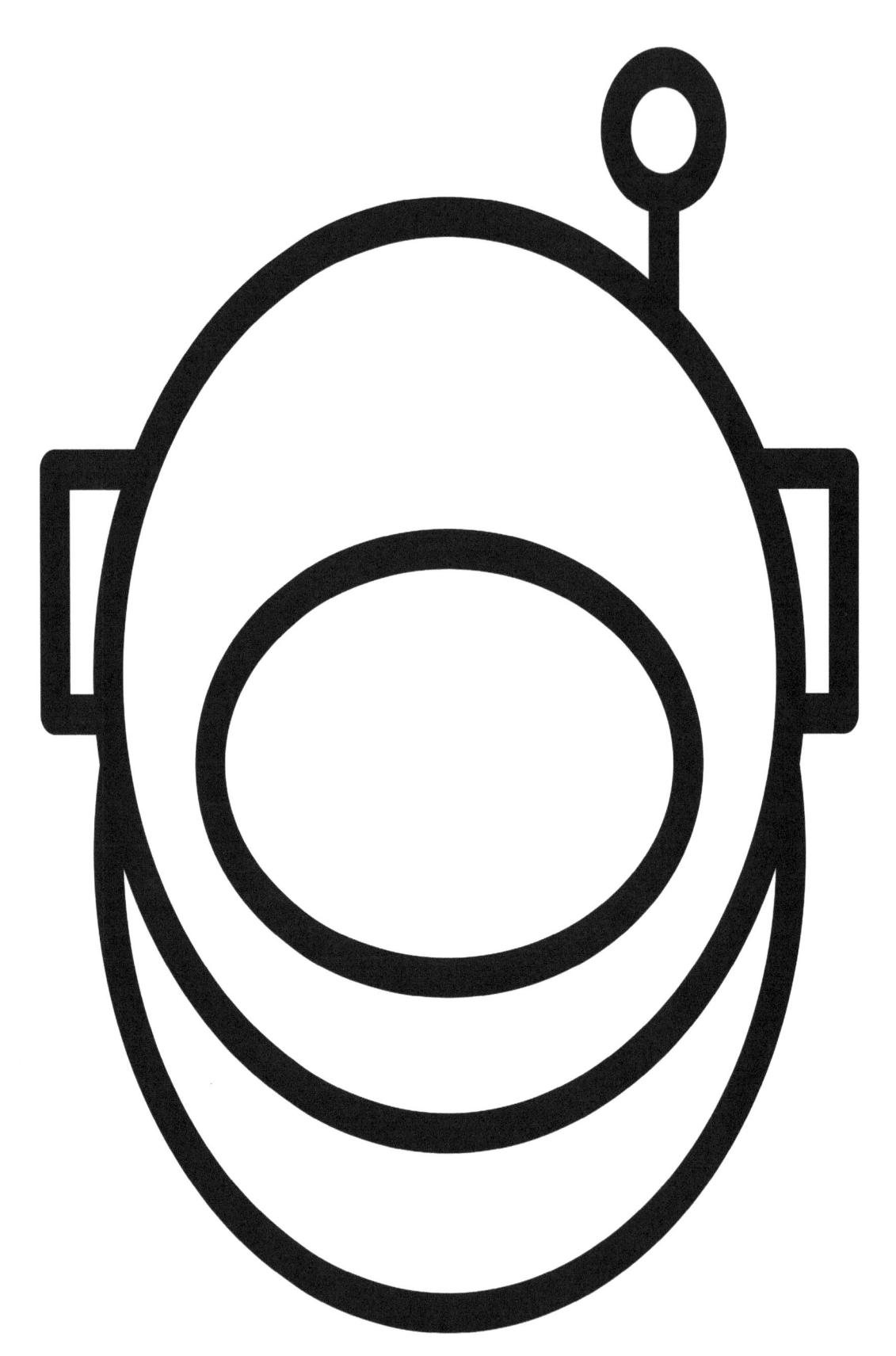

Draw the helmet

Count and color the shooting stars

Color the space icons

  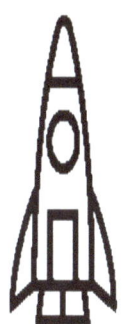 

 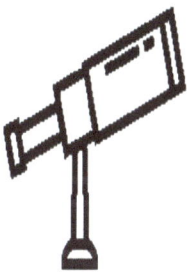  

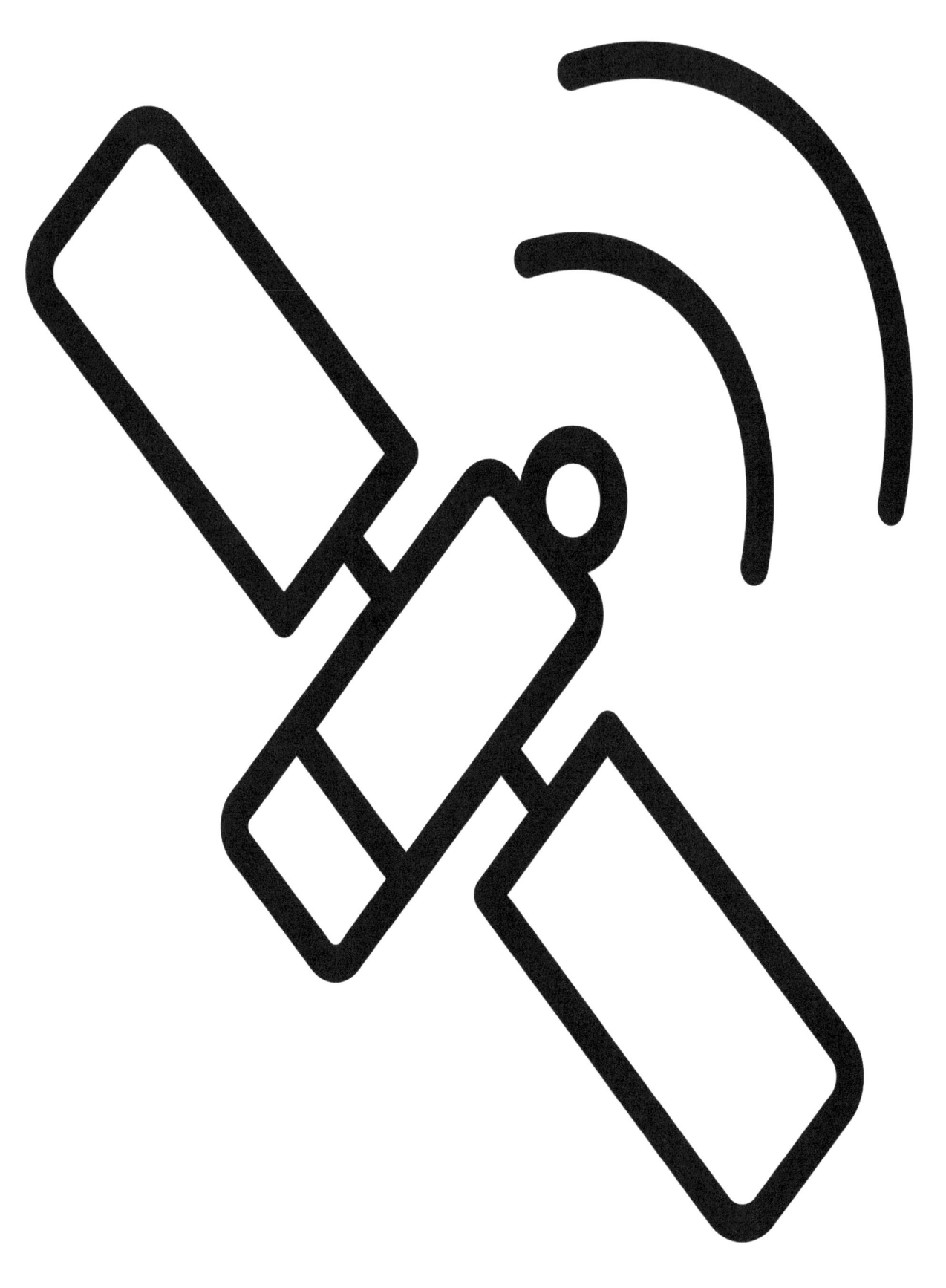

Draw the satellite

Draw the flag on the moon

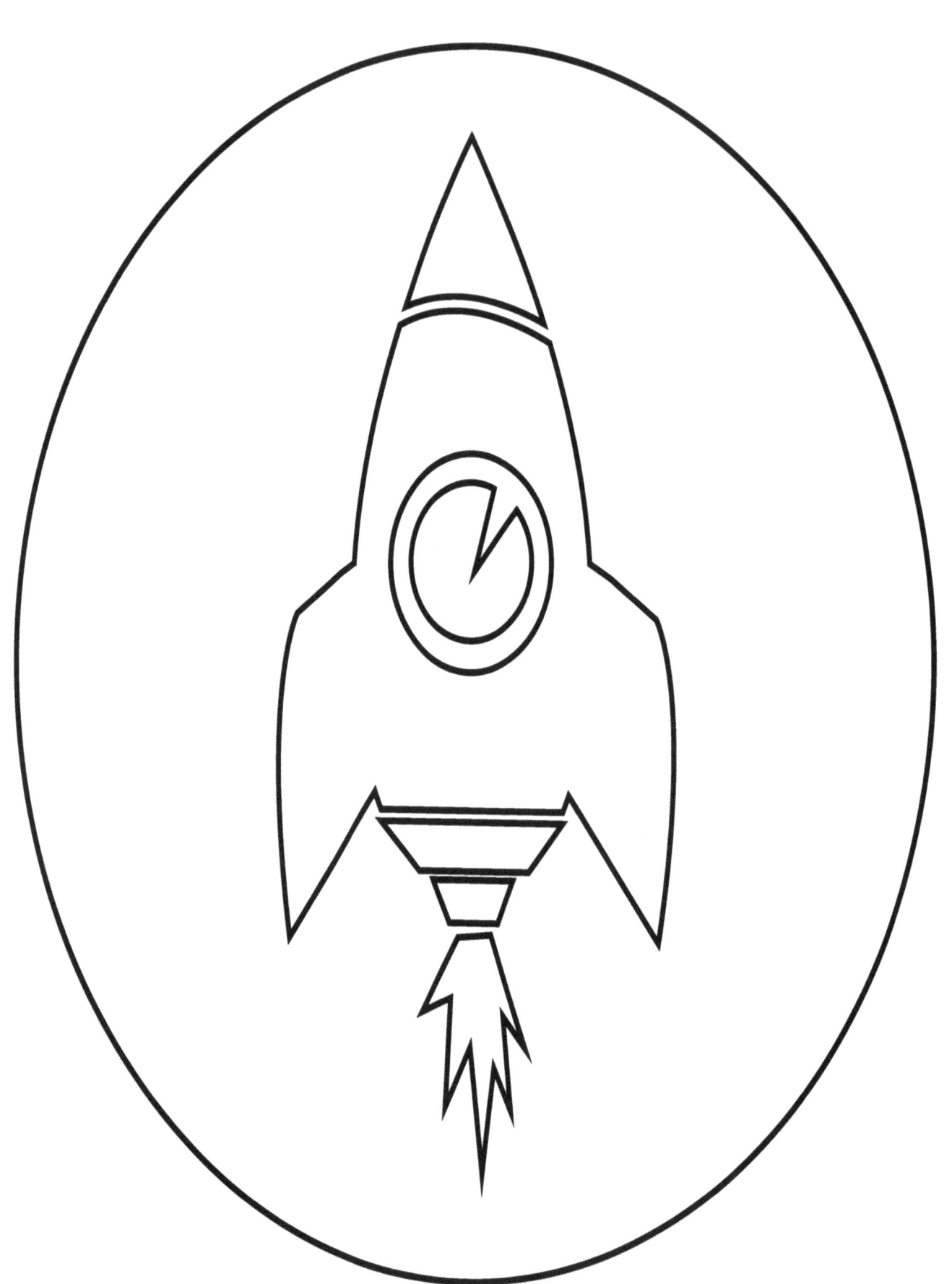

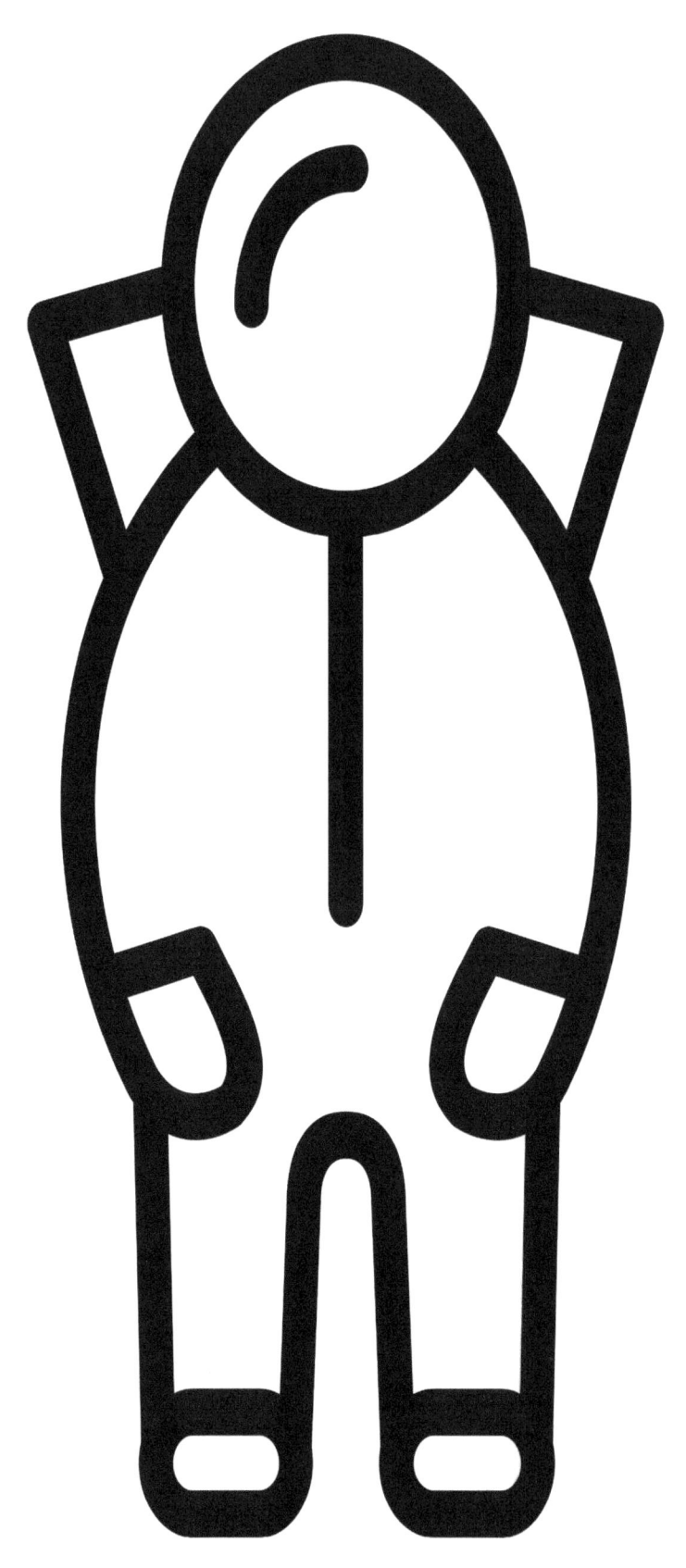

Write the words

LAUNCH _____

TAKEOFF _____

ROCKET _____

SPACE _____

HELMET _____

ASTRONUAT _____

SATALLITE _____

PLANETS _____

Draw the planet

THE END

www.ingramcontent.com/pod-product-compliance
Lightning Source LLC
Chambersburg PA
CBHW041936240526
45473CB00034B/1724